Awesome, Disgusting Science

GROSS SCIENCE OF PARASITES

Stephanie Bearce

BLACK RABBIT BOOKS

Hi Jinx is published by Black Rabbit Books
P.O. Box 227, Mankato, Minnesota, 56002.
www.blackrabbitbooks.com
Copyright © 2026 Black Rabbit Books

Alissa Thielges, editor; Jason Knudson, designer and photo researcher

All rights reserved. No part of this book may be reproduced in any form without written permission from the publisher.

Library of Congress Cataloging-in-Publication Data
Names: Bearce, Stephanie author
Title: Gross science of parasites / by Stephanie Bearce.
Description: Mankato, MN: Hi Jinx, an imprint of Black Rabbit Books, [2026] | Series: Awesome, disgusting science | Includes bibliographical references and index. | Audience: Ages 8-12 | Audience: Grades 4-6
Identifiers: LCCN 2025017489 (print) | LCCN 2025017490 (ebook) | ISBN 9781645824947 library binding | ISBN 9781645825005 paperback | ISBN 9781645825067 ebook
Subjects: LCSH: Parasites—Juvenile literature
Classification: LCC QL757 .B385 2026 (print) | LCC QL757 (ebook) | DDC 578.6/5—dc23/eng/20250702
LC record available at https://lccn.loc.gov/2025017489
LC ebook record available at https://lccn.loc.gov/2025017490

Printed in the United States of America.

Image Credits

Freepik/AI Image Creator, cover, 1, 13, almujaddidi, 17, brgfx, cover, 1, 5, 6, 20, freepik, cover, 1, macrovector_official, cover, 1, starline, 4, 5; Getty Images/Kateryna Kon/Science Photo Library, 2-3, 10-11; Science Source/Clinical Photography, Central Manchester University Hospitals NHS Foundation, 11; Shutterstock/Ak studio and photography, 7, Andrei Metelev, 11, ArtFamily, 4, Brett Hondow, 15, Consolidated News Photos, 10, HHelene, 8, ijimino, 18, Kateryna Kon, 2, 3, 9, 12, 13, 16, 17, 21, KateStudio, 7, Korrakit Pinsrisook, 12, Memo Angeles, 20, Negro Elkha, 4, Rost9, 4, 14-15, Sinhyu Photographer, 7, SkazovD, 9, Stephen Bonk, 15, Tomatheart, 12, Vinicius R. Souza, 14, 23, Zay Nyi Nyi, 8; Wikimedia Commons/Christian Gloor, 19.

Every effort has been made to contact copyright holders for material reproduced in this book. Any omissions will be rectified in subsequent printings if notice is given to the publisher.

CONTENTS

CHAPTER 1
Pests Gone Wild........5

CHAPTER 2
The Experiments.......6

CHAPTER 3
Get in on the Hi Jinx..20

Other Resources..........22

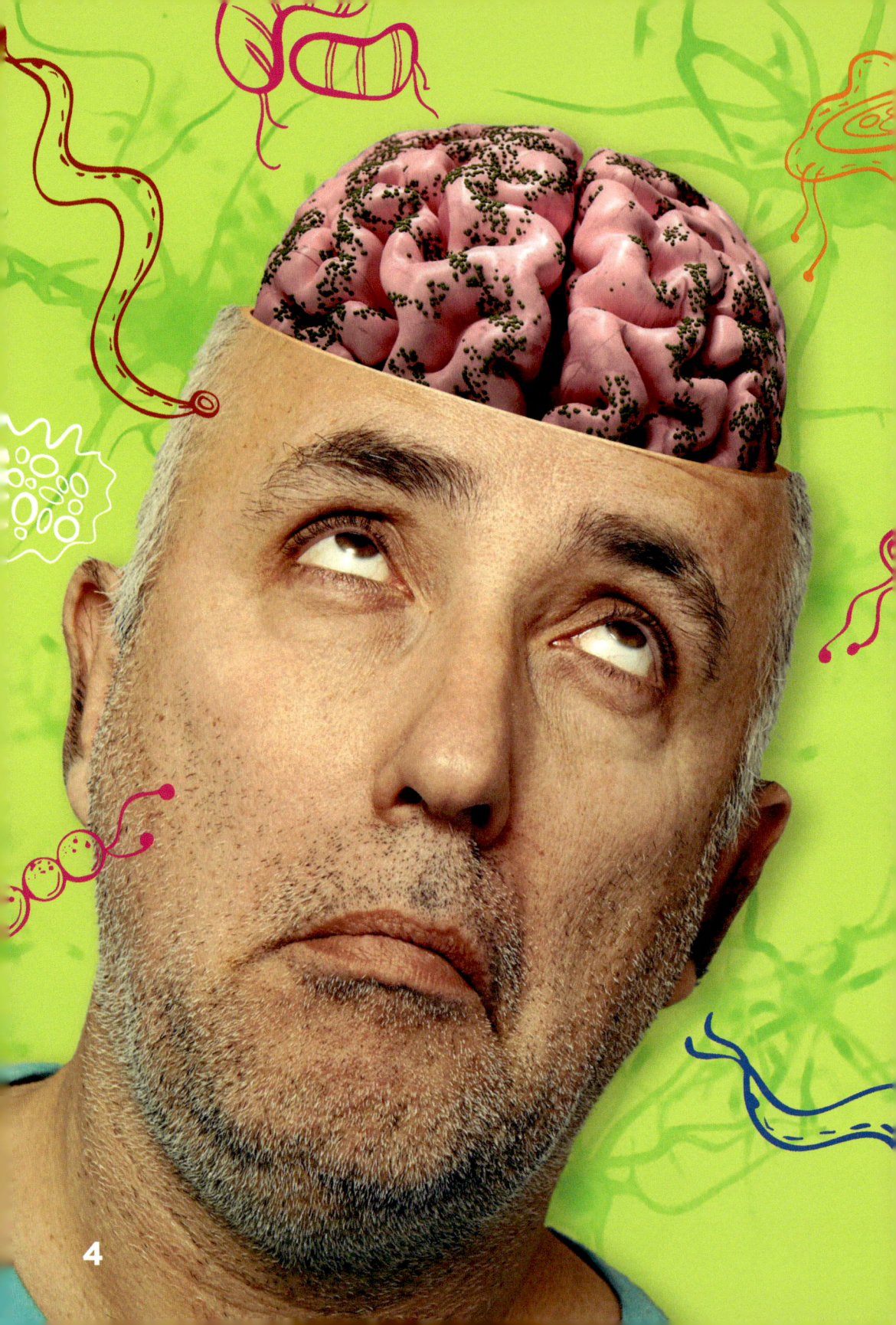

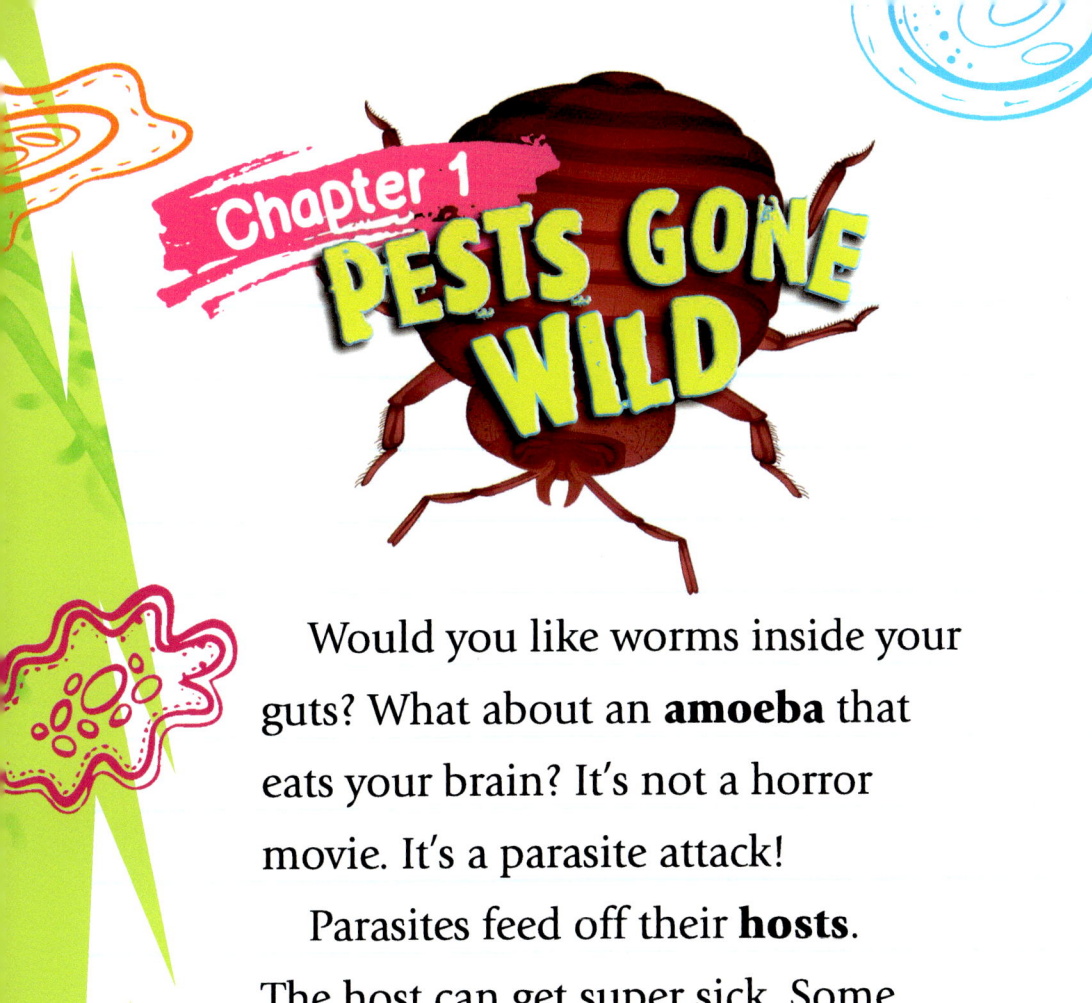

Chapter 1
PESTS GONE WILD

Would you like worms inside your guts? What about an **amoeba** that eats your brain? It's not a horror movie. It's a parasite attack!

Parasites feed off their **hosts**. The host can get super sick. Some hosts may die! Scientists study these pests. Their experiments can be gross—and awesome!

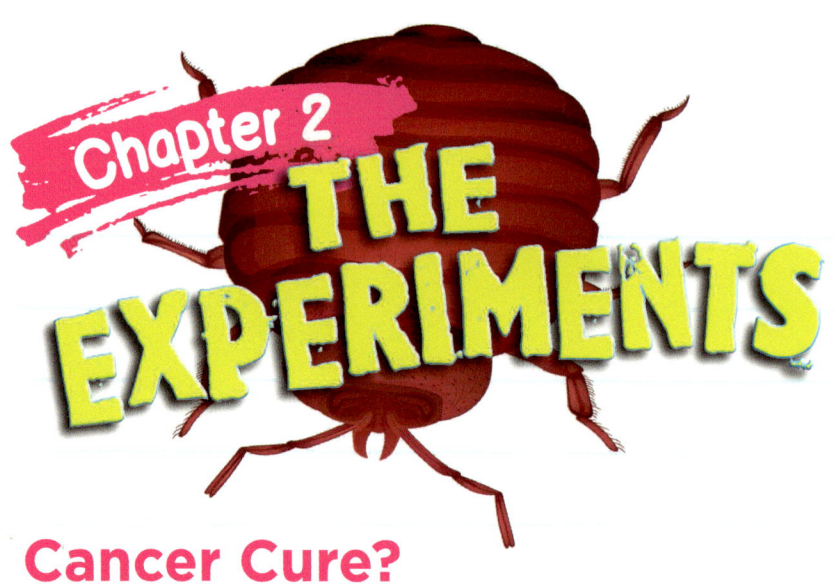

Chapter 2
THE EXPERIMENTS

Cancer Cure?

Tapeworms love human intestines. They are the perfect home. They're wet, warm, and full of food. But this is bad news for the host. Puss-filled lumps grow. The host gets **diarrhea**.

Tapeworms aren't all bad though. Scientists did experiments. These long worms make **molecules** that stop cancer cells from growing. Could this be a cancer cure? Maybe! More experiments are needed.

Tapeworms can grow to 30 feet (9 meters) long!

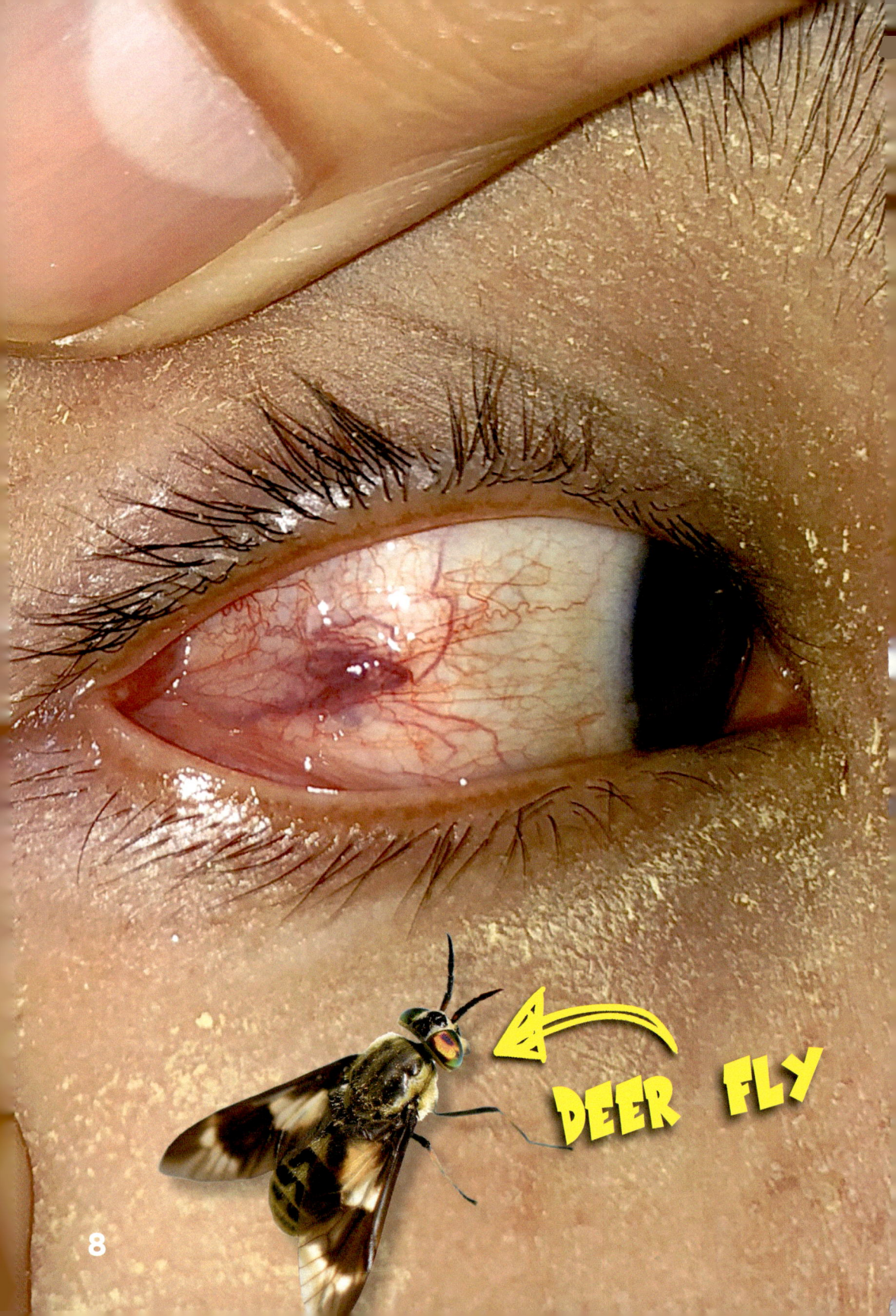

Eye Spy a Worm

The Loa loa is a nasty worm in Africa. It enters the body through deer fly bites. It travels through the bloodstream to the eyes. There it can cause blindness. Not good! Scientists want to stop this. They gave mice the worms. Then they tested different medicines. Doctors learned which ones worked the best. Now they can make medicine safe for humans.

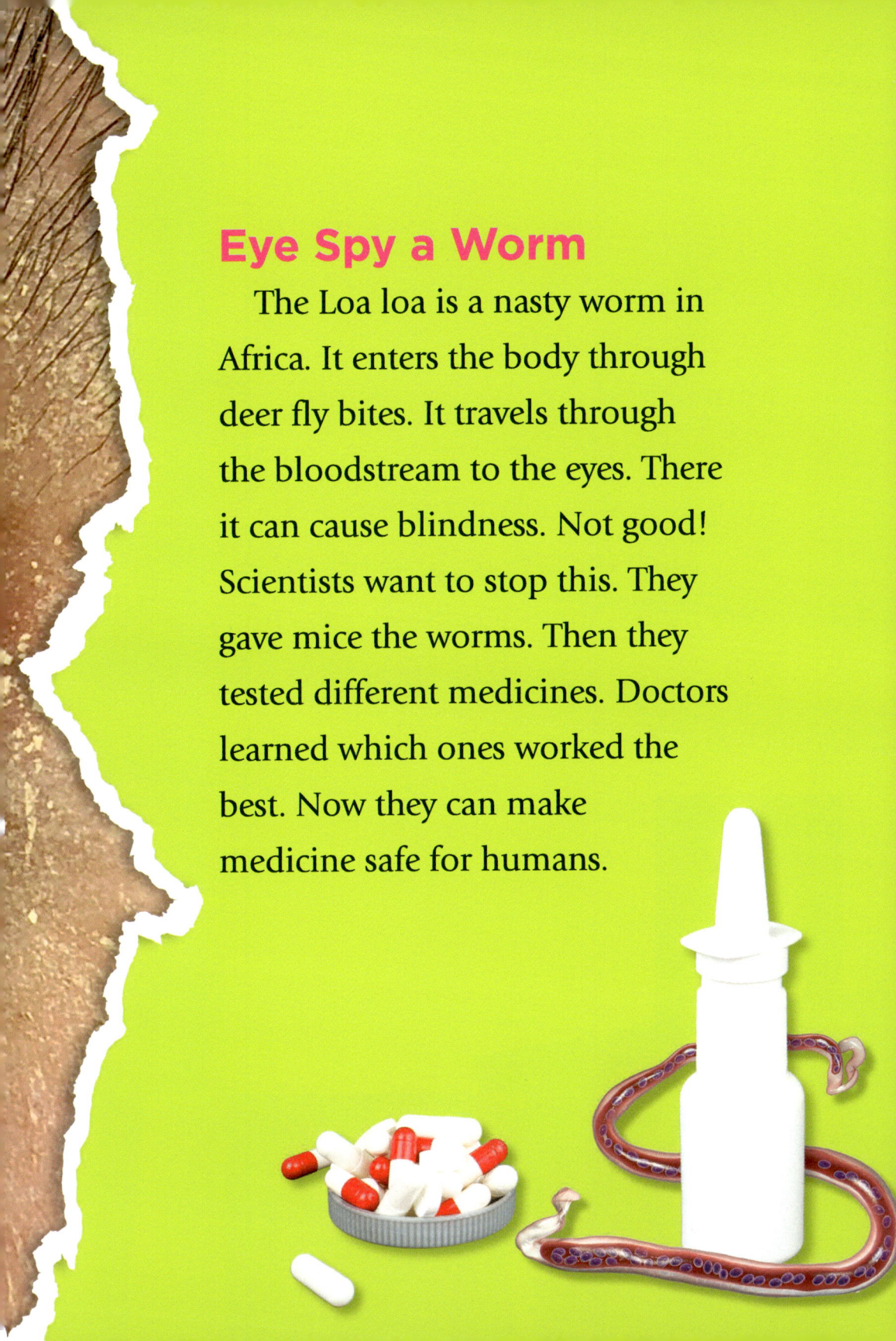

Worm Filters

Guinea worms grow in your gut. They can be 3 feet (0.9 meters) long. Burning blisters form. Then the worms crawl out! It's very painful.

Scientists learned the worms come from dirty water. Water **filters** help. They catch the **larvae**. In 1986, there were more than 3.5 million cases of guinea worm. In 2024, there were only 14 cases. Great work!

Former U.S. president Jimmy Carter helped stop this worm.

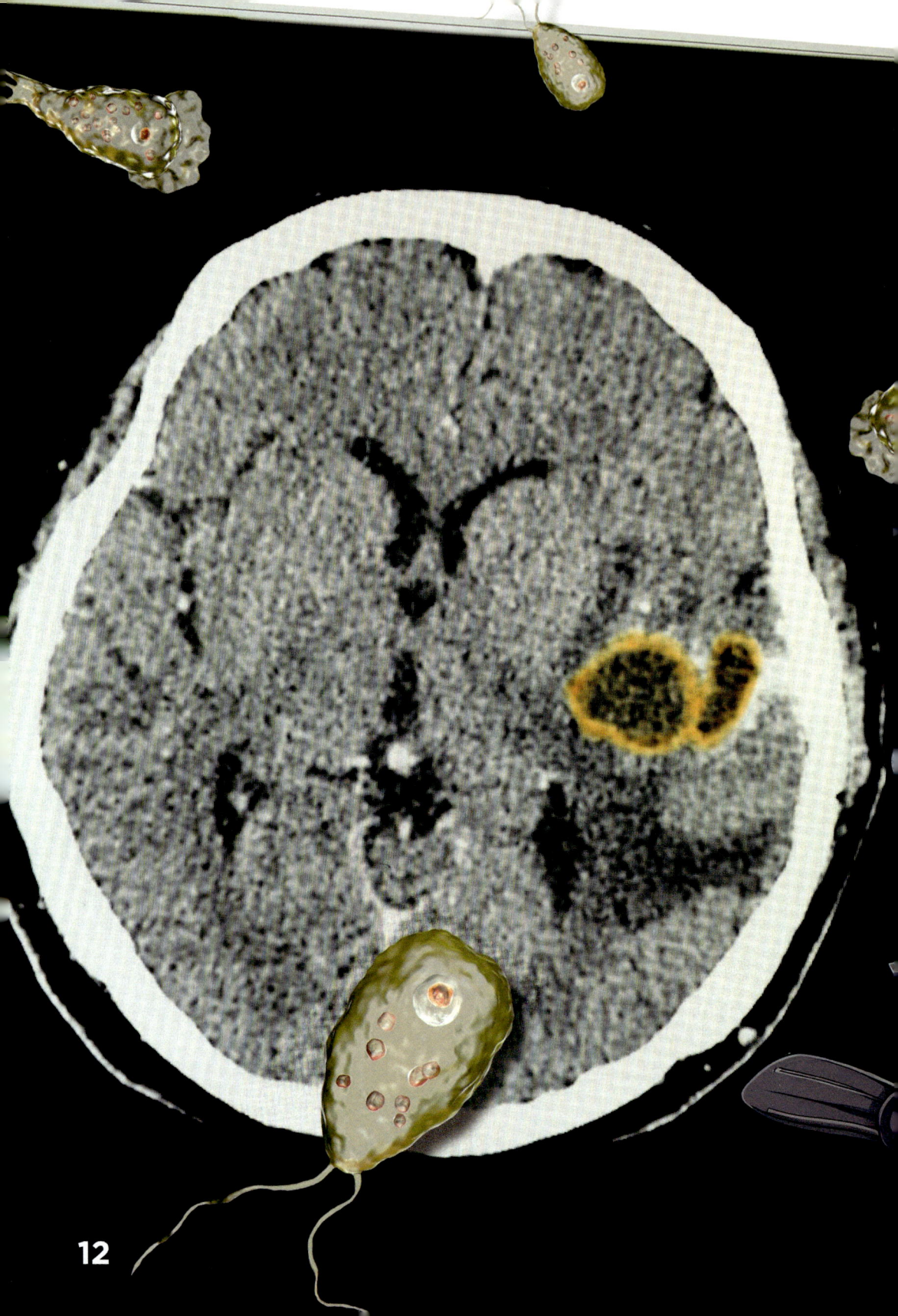

Brain Eaters

A tiny amoeba may lurk in warm fresh water. It swims up your nose. It munches on your brain. It could kill you!

Scientists did tests with these swimmers. They found a simple solution. Nose clips! Play it safe when swimming in warm water. Plug your nose!

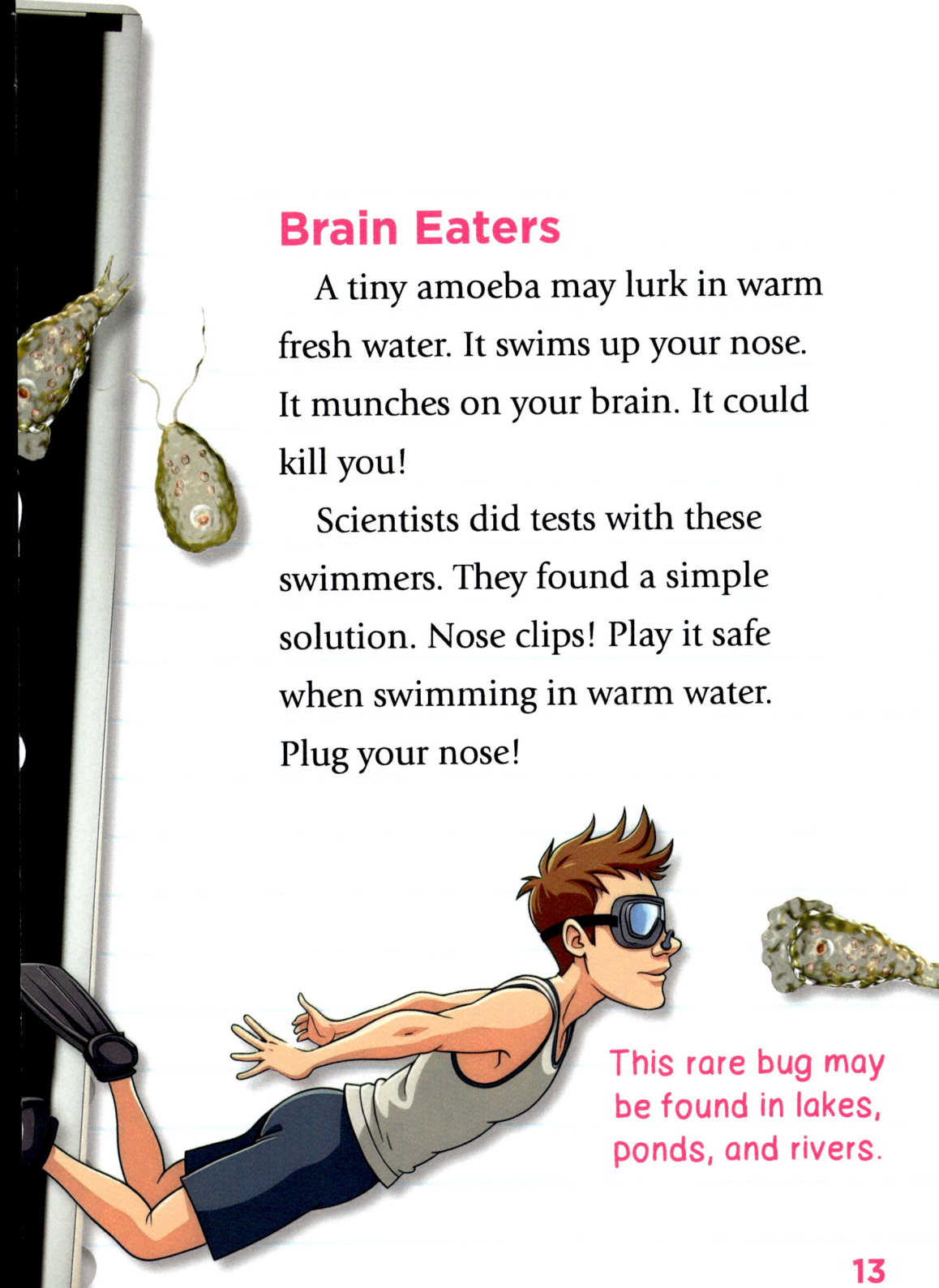

This rare bug may be found in lakes, ponds, and rivers.

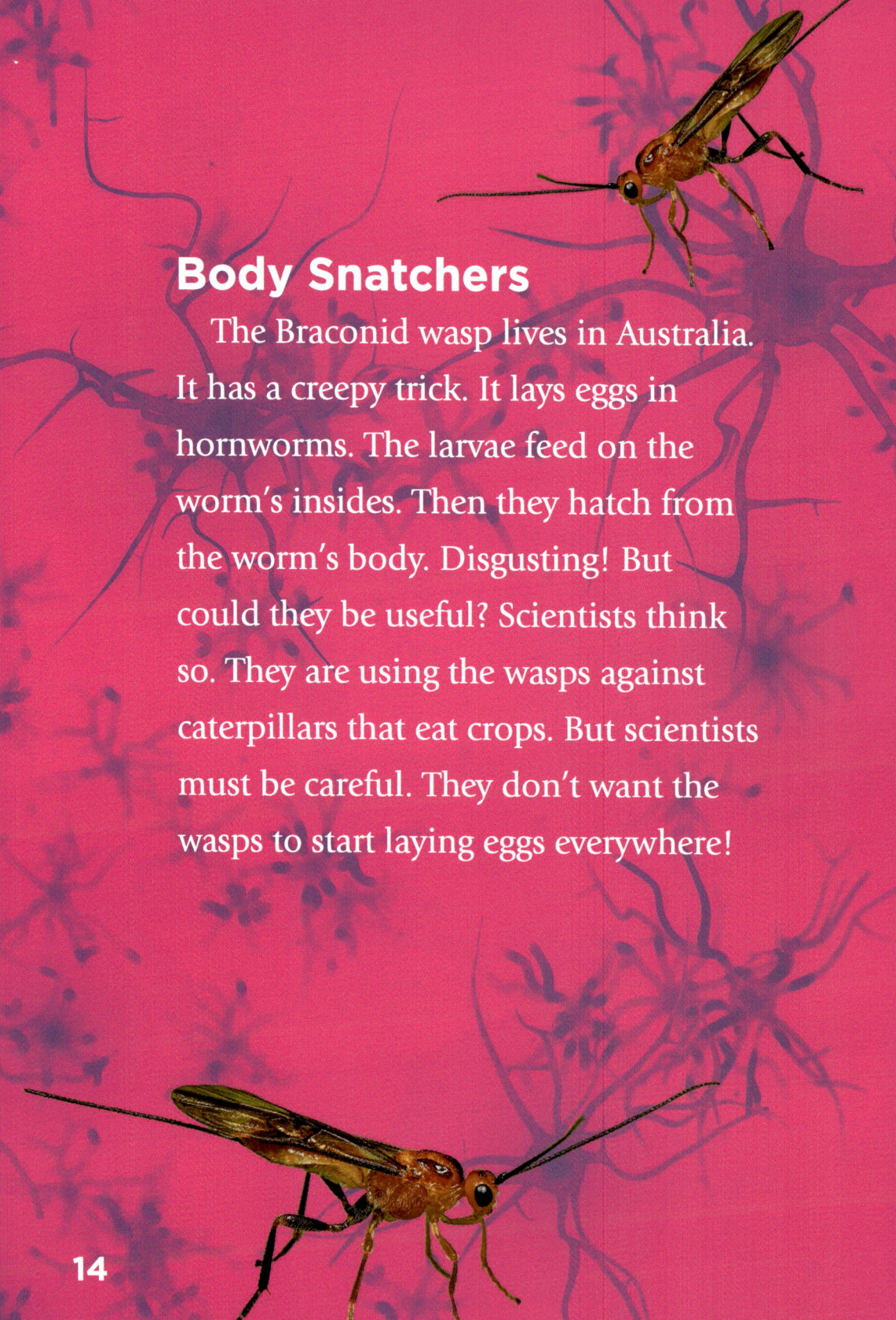

Body Snatchers

The Braconid wasp lives in Australia. It has a creepy trick. It lays eggs in hornworms. The larvae feed on the worm's insides. Then they hatch from the worm's body. Disgusting! But could they be useful? Scientists think so. They are using the wasps against caterpillars that eat crops. But scientists must be careful. They don't want the wasps to start laying eggs everywhere!

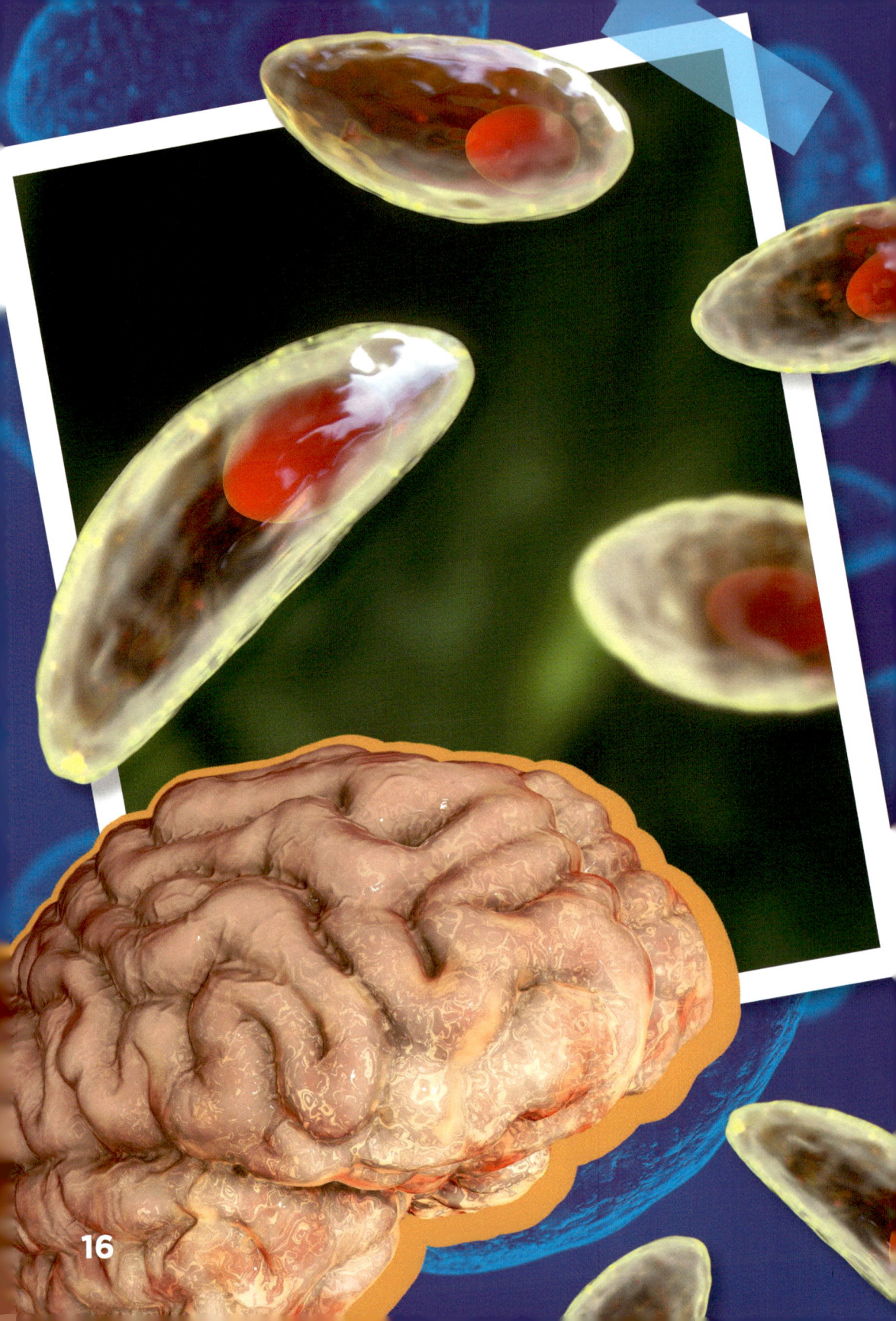

Mind Control

Some parasites hijack the brain. Scientists watched wild hyenas. Some were acting weird. Scientists nabbed those animals and took blood samples. Surprise! They found a parasite. It's called *Toxoplasma gondii*. The sick hyenas weren't scared of lions. The hyenas were easier to kill. That's what the parasite wanted. It can only **reproduce** in the body of a cat. Talk about freaky!

Tongue Stealers

One parasite steals fish tongues. Seriously! Scientists watched the host fish in the ocean. They learned a sneaky crustacean crawls in the fish's gills. It latches onto the fish's tongue. It sucks the blood until the tongue falls off. Then the parasite becomes the tongue! The fish can still eat with its freaky tongue. How gross is that?

The parasite is called *Ceratothoa springbok*.

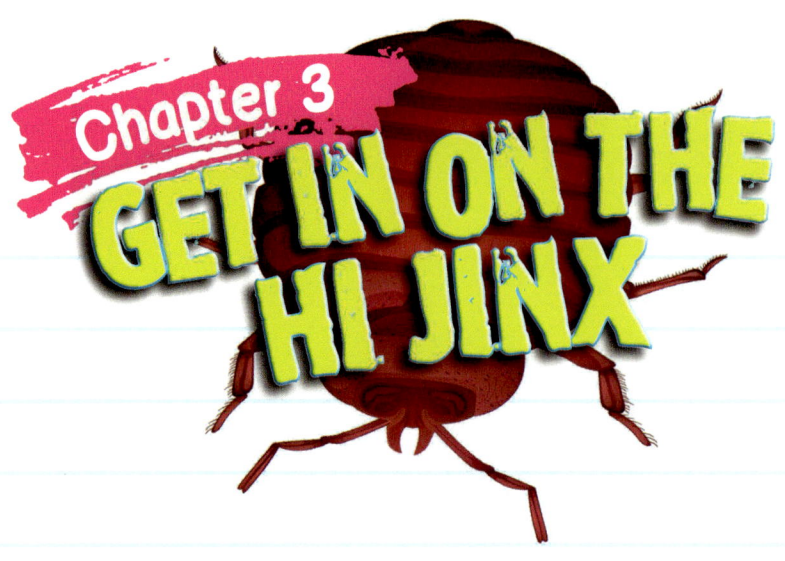

Chapter 3
GET IN ON THE HI JINX

Love animals and hate creepy crawlies? Be a vet tech! You'll help keep pets safe from nasty parasites. You'll give medicine and help spot signs of worms and bugs. Study biology and chemistry. Then find a vet tech program. You'll learn animal anatomy. Soon you'll be saving furry friends.

Take It One Step More

1. How do different parasites take over their hosts?

2. How does the Braconid wasp help its natural environment?

3. Do you think a parasite would ever be a good thing? Find examples in this book. Then search for more in other books.

GLOSSARY

amoeba (uh-MEE-buh)—a tiny living thing that is made of a single cell

diarrhea (dahy-uh-REE-uh)—an illness that causes you to poop frequently, usually in liquid rather than solid form

filter (FIL-tur)—a substance with pores through which a gas or liquid is passed in order to separate out floating matter

host (HOHST)—a living thing where an animal or plant lives and gets food or protection

larva (LAR-vuh)—the wormlike form of an animal that hatches from an egg

molecule (MOL-uh-kyool)—the smallest possible amount of a substance that has all the characteristics of that substance

reproduce (ree-pruh-DOOS)—to produce new individuals of the same kind

LEARN MORE

BOOKS

Lundgren, Julie K. *Gross and Disgusting Parasites.* New York: Crabtree Publishing, 2022.

Mather, Charis. *Painful Parasites.* Minneapolis: Bearport Publishing Company, 2024.

Wood, Chelsea L. *Power to the Parasites!* New York: Godwin Books, 2024.

WEBSITES

The Tongue-Eating Parasite
tpt.pbslearningmedia.org/resource/arct14.sci.nvtonguep/the-tongue-eating-parasite/

What Is a Parasite?
wonderopolis.org/wonder/what-is-a-parasite

INDEX

A
amoebas, 5, 13

B
blindness, 9

C
cancer, 6
caterpillars, 14

D
diarrhea, 6

F
fish, 18

H
hyenas, 17

S
swimming, 13

V
vet techs, 20

W
wasps, 14, 21
water filters, 10
worms, 5, 6, 7, 9, 10, 14, 20